Diagnosing Bee Mites

D. Sammataro, PhD

Entomologist/ Beekeeper

USDA-ARS (Retired)

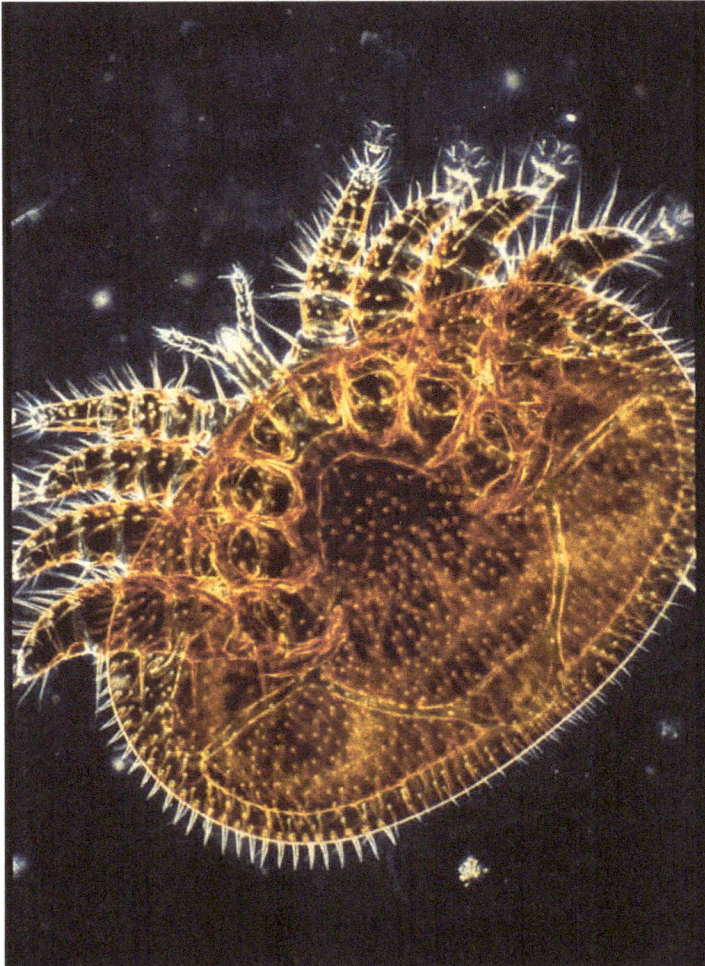

Diagnosing Bee Mites

© D. Sammataro, PhD

ISBN 978-1-908904-59-1

Published by Northern Bee Books, 2014
Scout Bottom Farm
Mytholmroyd
Hebden Bridge HX7 5JS (UK)

Photographs by D. Sammataro unless otherwise credited

Design and Artwork, D&P Design and Print

Printed by Lightning Source UK

Diagnosing Bee Mites

D. Sammataro, PhD

Entomologist/ Beekeeper

USDA-ARS (Retired)

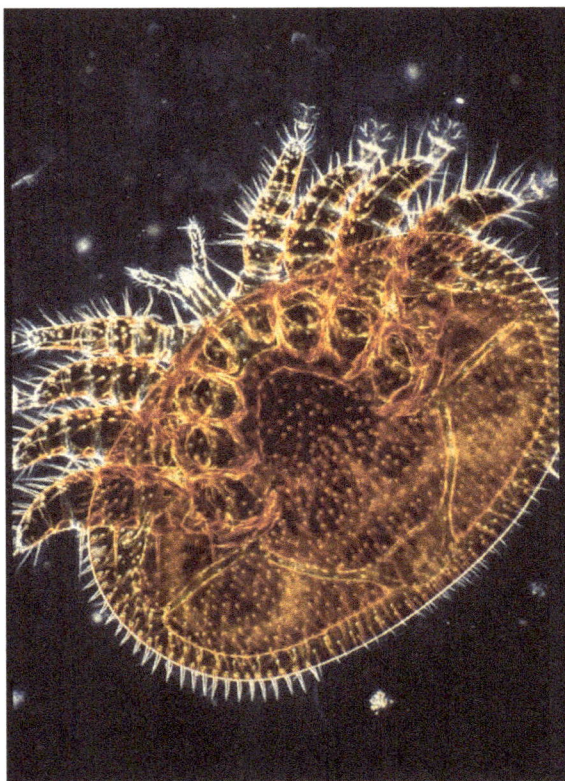

NORTHERN BEE BOOKS

Table of Contents

Written/produced by D. Sammataro,
Diana Brand Honey Bee Research Services,
2014 © dsammbeegirl@gmail.com

Varroa Mites on Honey Bees

Family: Varroidae (*Delfinado & Baker, 1974*)

Source: Mites of the Honey Bees

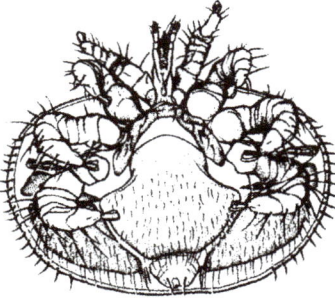

Genus: *Varroa*

Varroa jacobosoni Oudemans 1904 (on Asian honey bees *A. cerana, A. nuluensis*)

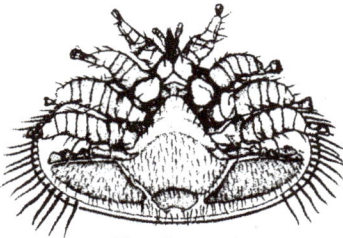

V. underwoodi (on Asian honey bees: *Apis cerana, A. nigrocincta*, A. florea)

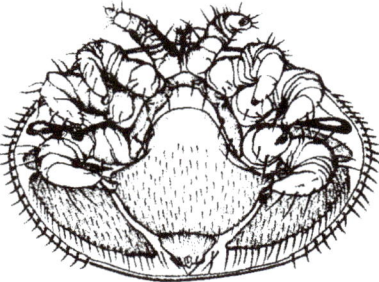

V. rindereri (on *A. koschevnikovi*)

V. destructor Anderson & Trueman 2000 (on *A. cerana, A. mellifera, A. m. scutellata, A. dorsata dorsata*) and where European honey bees have been introduced

Photo by D. Sammataro

I

Size Comparisons of Bee Mites with a Flea

All photos are approximately the same scale

Flea on dog

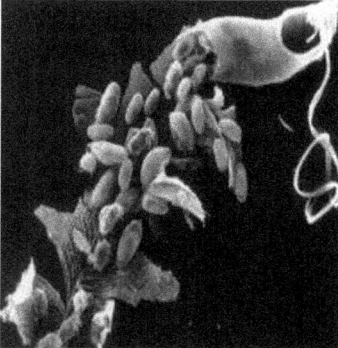

Tracheal mites inside an opened tracheal tube

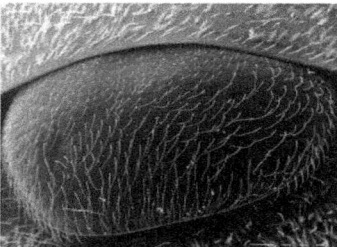

Varroa mite. The mite is hiding under the sclerite of a bee

Photos by W. Styer, Ohio State University, Wooster

Healthy Brood Frames

• Bee population high

• Brood pattern solid

• Mites not readily visible

Signs of a Colony in Distress

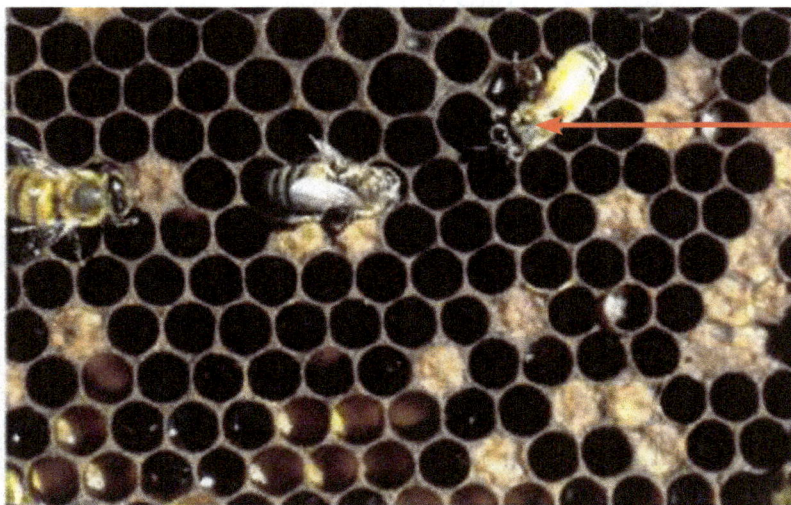

- Bee population usually low

- Brood pattern is spotty and cappings have sunken appearance

- Brood (when opened) and young bees will have visible signs of mites (arrow)

- **Make sure there are no other disease problems; such as foulbrood or chalkbrood diseases**

Visible Symptoms of Mites on Bees

Varroa on emerging bee (arrow)

Virus symptoms: Deformed Wing Virus (DFW)
(wings crumpled, young bees small and weak),

USDA photographs

Varroa on Chalkbrood Mummies

Varroa can be also found on Chalkbrood mummies and affected larvae.

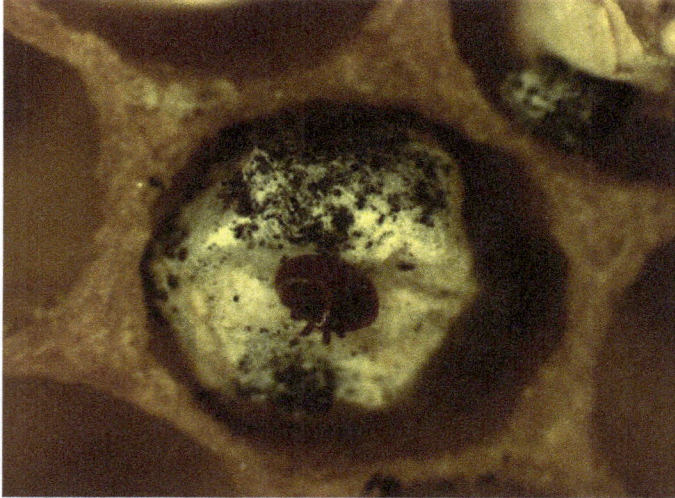

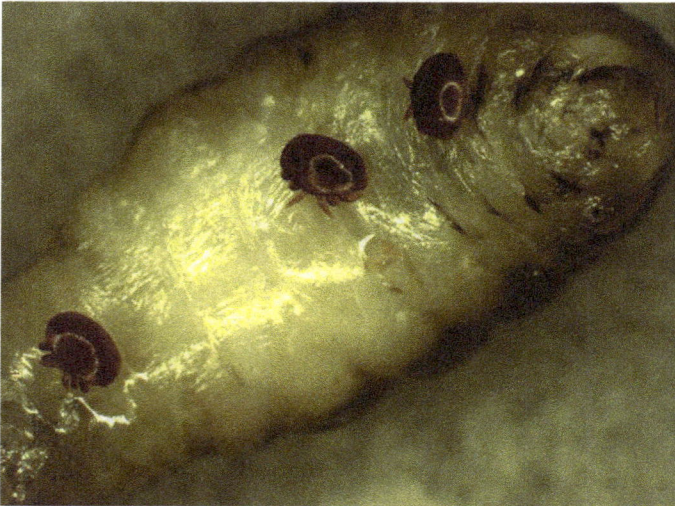

Sammataro, D. and J. Finley, 2004 (see References)

Life Cycle of Varroa

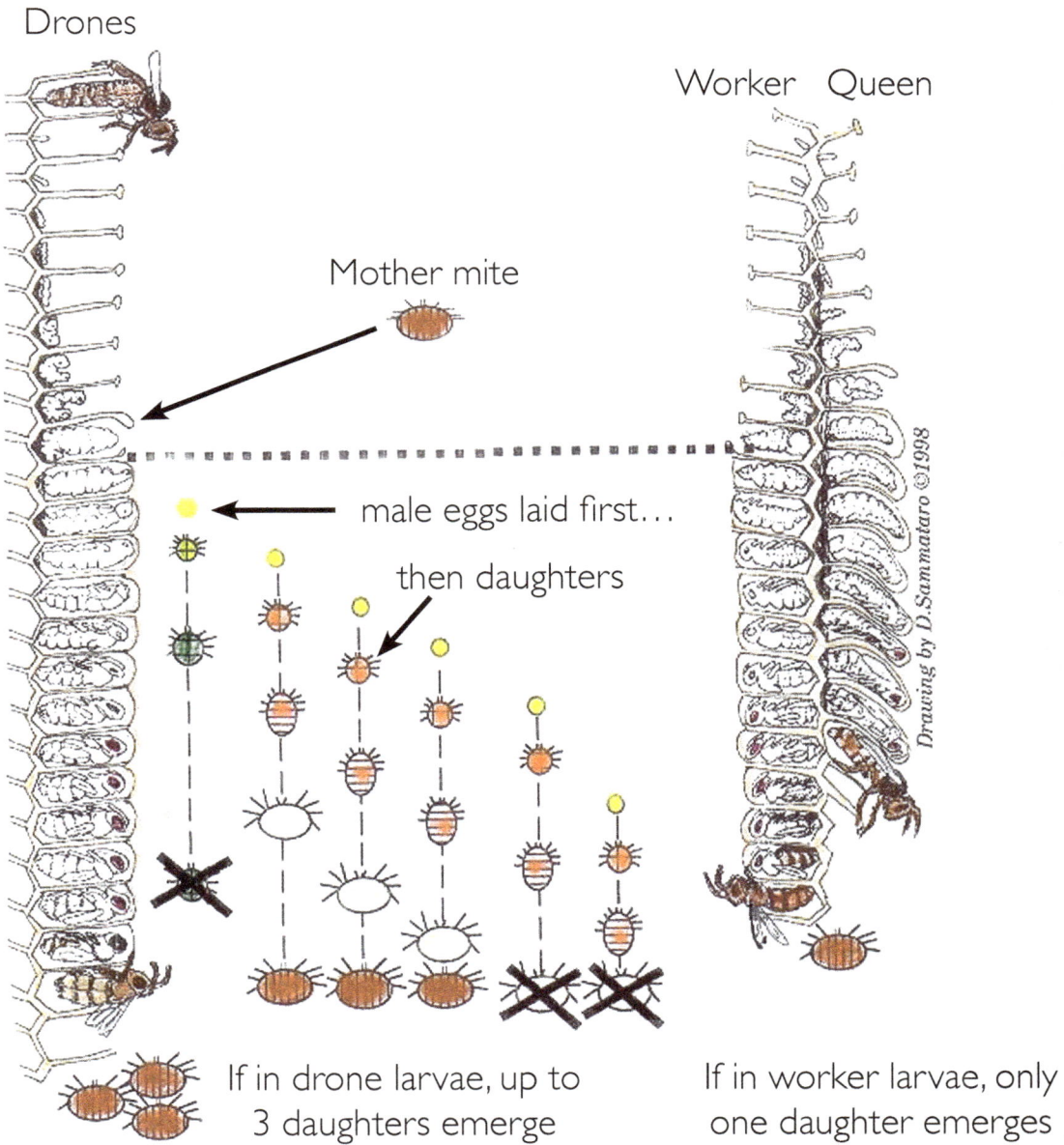

Drones

Worker   Queen

Mother mite

male eggs laid first...

then daughters

Drawing by D.Sammataro ©1998

If in drone larvae, up to
3 daughters emerge

If in worker larvae, only
one daughter emerges

Current Range of Varroa

This is a global parasite, spread by migrating bee colonies (moved by beekeepers).

An animated map of the spread of the Varroa mite can be found at: *http://www.mylovedone.com/image/solstice/win12/ varroa2012b. html*

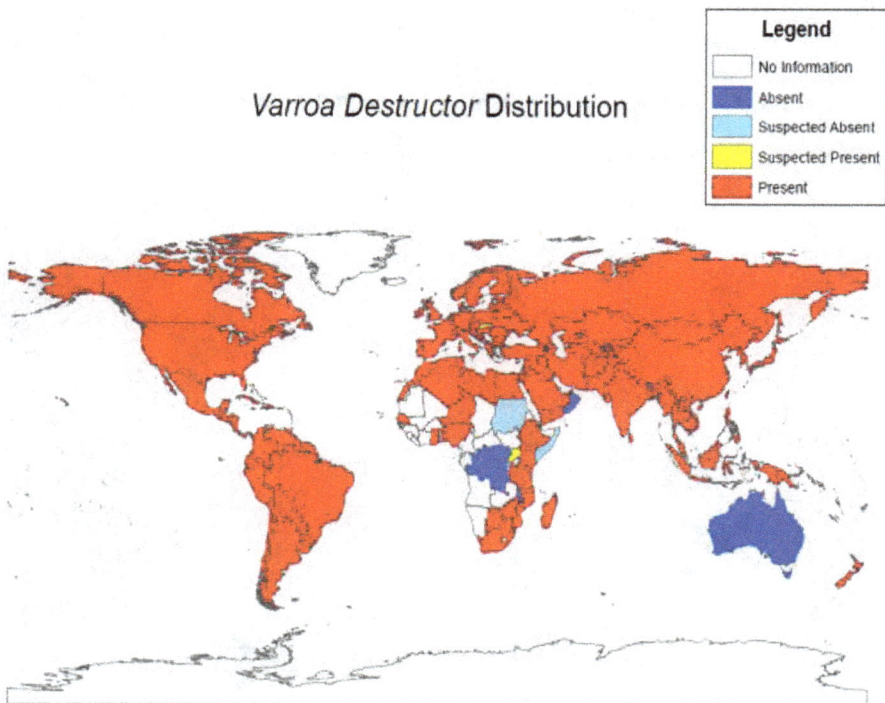

Varroa Destructor Distribution

Legend	
	No Information
	Absent
	Suspected Absent
	Suspected Present
	Present

Ellis and Munn, 2005 and Ellis, J. et al. www.ufhoneybee.com 2014 (see References)

Three ways to Diagnose Varroa (also *Tropilaelaps*) Mites

M. Frazier

1. Ether or Sugar Roll (wash)

Sammataro

2. Uncapping drone brood

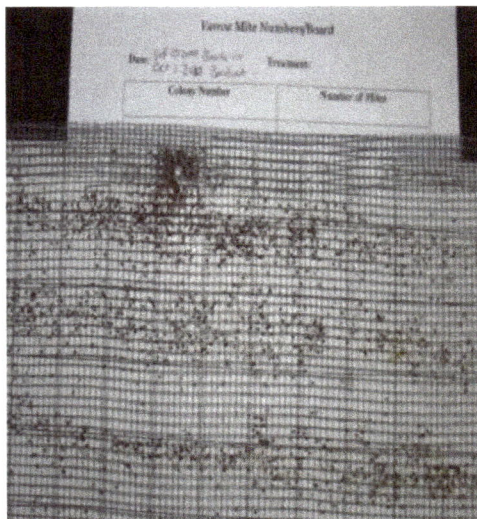

3. Sticky boards or bottom debris

Sampling Methods

1. Rolling or shaking bees, using:

   A. *Ether*

   B. *Soapy Water or alcohol*

   C. *Sugar Shake*

2. Brood examination

3. Hive debris/sticky boards

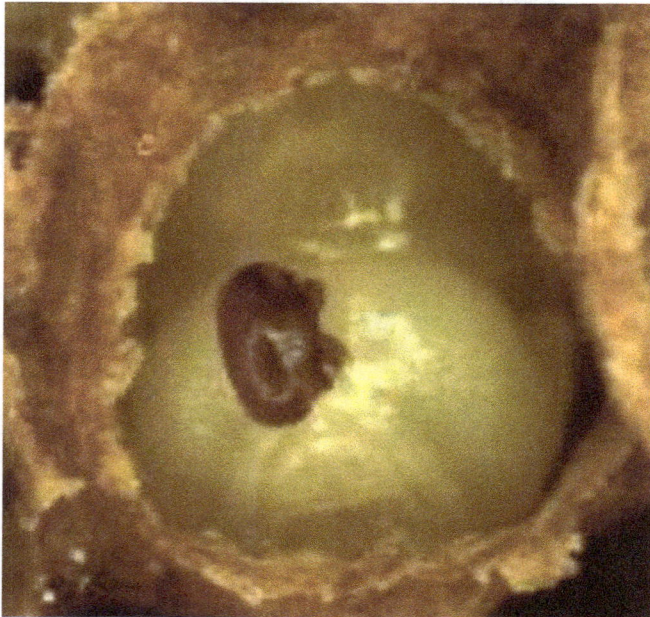

1A. Ether Roll

Brush or shake 300 bees into a wide-mouth jar from one frame with capped brood and nurse bees. **NOT THE QUEEN!!!**

300 bees = 2 inch (5.08 cm) layer of bees

- 100ml (0.42 cup) will hold 300 bees

- make your own marked container

1/3 cup (78.07 milliliters) is about 238 bees

2 Inches

1A. Ether Roll

1. Apply a two second burst of ether (automotive starter fluid) into the jar.
   (**FLAMMABLE** - keep away from fire and working smokers).

2. Replace the jar top and shake vigorously for 30 seconds.

3. Gradually rotate the jar horizontally.

4. Look for any mites sticking to the sides; ether kills the bees and causes them to regurgitate, making the jar sides sticky.

Photos by D. Sammataro

1B. Soapy Water Wash

1. Scrape 300 bees into jar, again *make sure* that the **QUEEN IS NOT INCLUDED!**

2. Add 50 ml (~ 2 oz. or 1/3 cup) of wind-shield wiper fluid, rubbing alcohol or *soapy water*, shake vigorously for several minutes.

3. Remove jar lid and pour contents through metal wire-mesh screen.

4. Repeat several times on same 300 bees.

5. Count mites by pouring mite/water through a coffee filter.

Photo: www.extension.umn.edu/honeybees

1C. Sugar Shake

Photos D. Sammataro

Collect 300 live hive bees (not queen) from brood frames
(2 inches in quart jar or 100mL beaker = 300 bees)

Add the
300 bees
into jar with
a screened
mesh lid;
(cut wire
mesh to fit
into jar ring;
arrow)

Add Powder:
Confectioner's Sugar or flour

Add 2 Tablespoons (25 g or 1 oz.) of:

- Powdered Sugar (also called Icing sugar)

  or

- Ground white sugar (or Flour)

1C. Sugar Shake (Cont'd)

Rest for 1 minute, then:

▶ Turn jar over (wire screen down) and vigorously shake jar like a salt shaker.

▶ Shake over a white surface.

▶ Repeat the above two steps until virtually no more sugar shakes out.

Mason jars with wire mesh strainer

Photos D. Sammataro

1C. Sugar Shake: Counting Mites

To visualize mites:

▶ Shake onto white paper and spray mites with water,

or

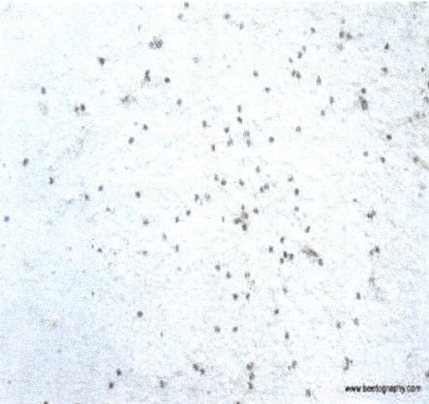

▶ Use paper towel and spray with water, or

▶ Shake mites into pan of water- mites will float;

- Strain through a coffee filter and count mites.

Photos D. Sammataro

Return Bees To Colony

Photos: D. Sammataro

Estimating Mite Density Numbers: Sugar Shake

▶ Estimate the density of mites (per 100 bees) in individual colonies:

- Collect ~300 adult bees from a single frame in upper broodbox to estimate colony mite density (sugar shake or ether roll)

▶ To figure colony mite density:

- Divide number of counted mites by 3 (to get mites per 100 bees) and multiply by a correction factor of 2 (to count the mites hidden in the brood).

  e.g. 12 mites on 300 bees (12/3=4) × 2= **8% infestation**

▶ For greater precision examine three, 300-bee sample units.

Lee et al. 2010a (See References)

Alternative way to Determine Mite Density: using this chart

If 12 mites were counted on 300 bees, check that number on the chart: 12 mites means an 8% total colony infestation level (mite density)

#Mites per 300 adult bees	Colony infestation	#Mites per 8 300 adult bee samples	Apiary infestation
1	1%	8	1%
2	1%	16	1%
3	2%	24	2%
4	3%	32	3%
5	3%	40	3%
6	4%	48	4%
7	5%	56	5%
8	5%	64	5%
9	6%	72	6%
10	7%	80	7%
11	7%	88	7%
12	8%	96	8%
13	9%	104	9%
14	9%	112	9%
15	10%	120	10%
16	11%	128	11%
17	11%	136	11%
18	12%	144	12%

Lee, K.V., G. Reuter and M. Spivak, 2010b (See References)

Sampling Apiaries

▶ For sampling whole apiaries, examine 300 adult bees from one brood frame in each of 8 randomly-selected colonies in each apiary, count mites.

▶ Check total mite numbers against the chart.

AZ apiary, Sammataro photo

Apiary Samples

So…if a total of 130 mites were counted from 8 colonies…

#Mites per 300 adult bees	Colony infestation	#Mites per 8 300 adult bee samples	Apiary infestation
1	1%	8	1%
2	1%	16	1%
3	2%	24	2%
4	3%	32	3%
5	3%	40	3%
6	4%	48	4%
7	5%	56	5%
8	5%	64	5%
9	6%	72	6%
10	7%	80	7%
11	7%	88	7%
12	8%	96	8%
13	9%	104	9%
14	9%	112	9%
15	10%	120	10%
16	11%	128	11%
17	11%	136	11%
18	12%	144	12%

…the estimated mite load is **11%** in this apiary

2. Brood Examination

▶ Up to 85% of the mites in a colony are in capped brood cells and therefore, not visually detectable.

▶ Varroa mites are more attracted to drone brood than worker brood, so look there first.

▶ Sample 100 – 250 cells.

▶ Locate a patch of drone cells that are in the purple-eye pupae stage (test open some brood); worker brood should also be examined.

▶ Slide the prongs of an uncapping fork along the comb face and into the cappings.

▶ Pry upward and remove the pupae.

▶ Carefully examine the bodies and the interior of the cells for mites.

Credit: Volcano Island honey company website.

2. Pulling up Brood

Good method **but**: no standard procedures for estimating mite populations…

Photo credit: Iowa Honey Producers

Examine brood for mites:

Worker pupae, and…

…drone pupae pulled up with a cappings scratcher.

Mites attached to the brood (arrows)

Photos: middle and bottom by A. Frake.

3. Sticky Boards (or Hive Debris)

Insert Sticky Board for 1-3 days.

Note: Commercial sticky boards are available from bee supply companies.

Homemade boards can be made from stiff paper or plastic covered with insect barrier or glue (e.g. Tanglefoot).

Cover boards with wire mesh to keep out bees (arrow); or use a screened bottom board.

Screened Bottom Boards

http://nhhoneybee.com/
bottomboardscreened.aspx

Making Screens for Sticky Boards

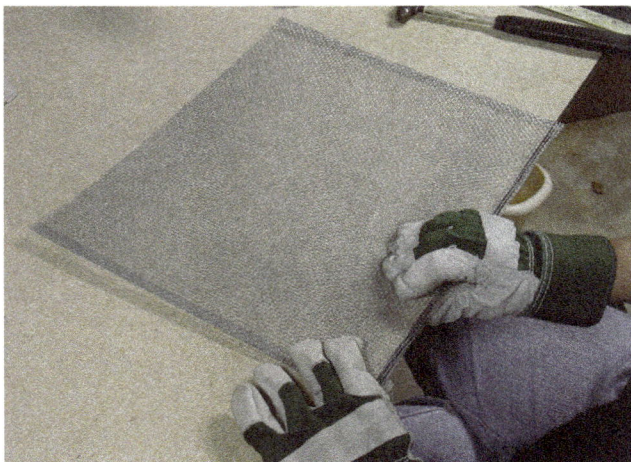

Use metal hardware cloth mesh; it must be washed first, otherwise the coating can kill bees. Cut the mesh to fit just inside the sticky coating, fold over edges.

Staple the wire screening to the board to keep bees from contacting the sticky material.

After 3 Days, Remove the Boards and Count Mites

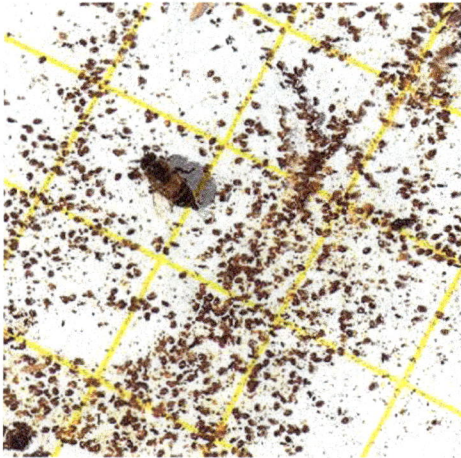

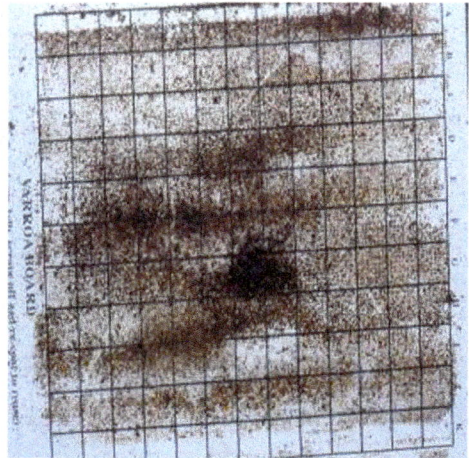

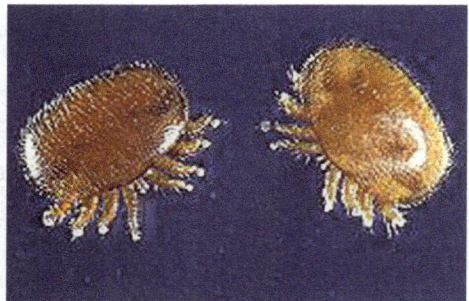

Mites are oval and hard to the touch

Photos: D. Sammataro **except mites, which are USDA**

Sticky Boards are also used to monitor the Asian *Tropliaelaps* mite

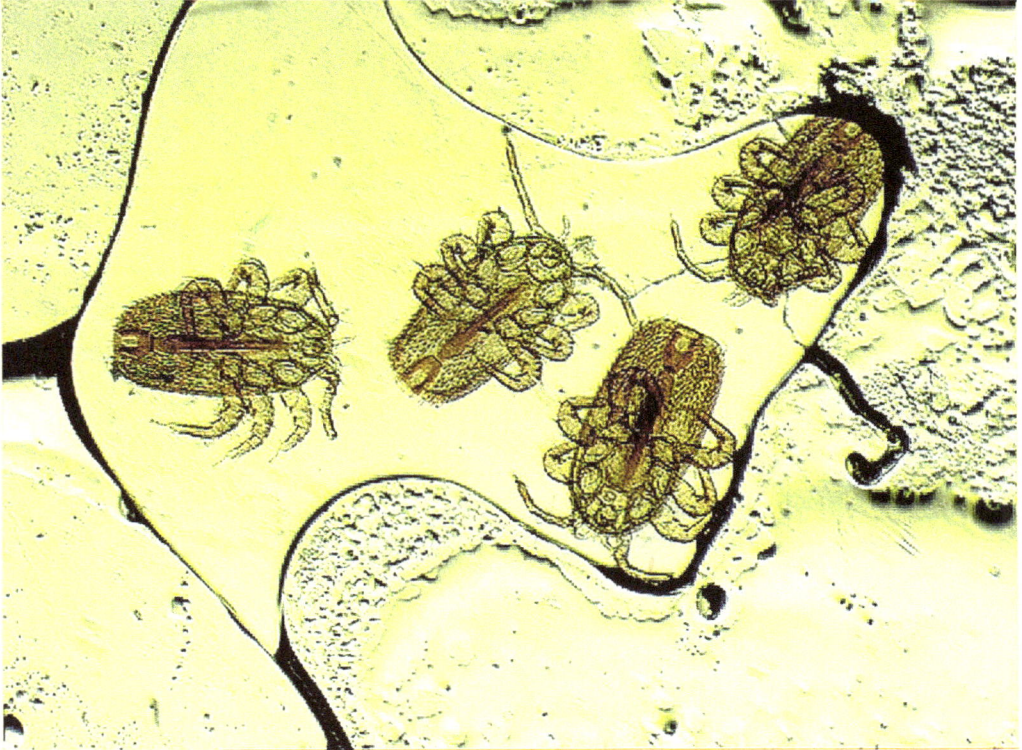

Photo by D. Sammataro

▶ Another Asian bee mite is on the horizon.

▶ Use the same sampling techniques to look for this mite (see page 53 for more information on *Tropilaelaps*).

Record Keeping is Important

DATE:_____

Yard:_____

Sticky Boards

Col. No.	Date IN	Date OUT	Mite numbers	Treatmnt Date	NOTES: (Chalk?, dead bees or pupae?, board chewed?)

Treatment Thresholds for Sticky Boards

▶ On a 24-hour sticky board in Northern US:

- 12 mites (spring)

- 23 mites (fall)

▶ In southeastern US states:

- 1-12 mites (spring)

- 71-224 (fall)

This means that if you count this many mites, you should start treating all colonies.

HOWEVER, this is a rough estimate.

It is important to track changes in mite density in colonies over time.

Delaplane, *et al.* 2005 (see References)

Treatment Thresholds for Rolls and Shakes

(Northern regions)

▶ Controls will be needed if more than 10-12% adult bees are infested

    For stationary colonies, not for migratory operations

▶ If less than 10%, no treatment needed

▶ If threshold is just under 10%, this is a gray area, could treat or not treat

    ✓ Keep sampling, **monitor changes mite densities over time**

Factors Influencing Mite Populations

▶ Time of year

▶ Colony size

▶ Hygienic behavior of bees (bees pulling out dead or diseased bees)

▶ Amount of drone brood

   • If amount of drone brood is small, mites will invade worker brood

   • Only a small percentage of mites could be in drones

When to Test (or Monitor) Mite Densities

▸ Early detection offers the best opportunity for effective Varroa control.

▸ Frequency of colony testing:

　　1. If mite density is low, twice a year.

　　2. If mite density is high, every two months.

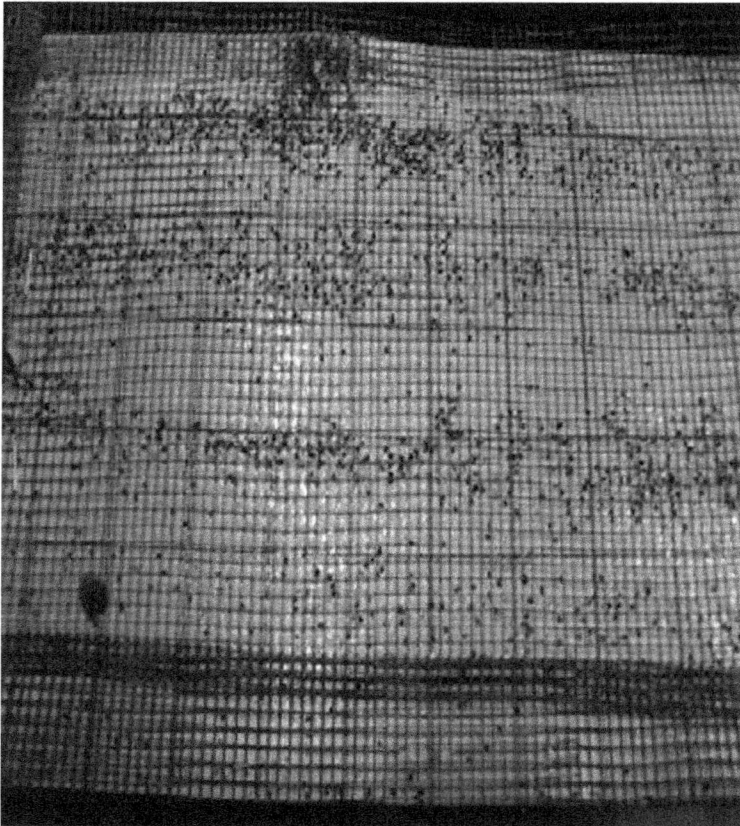

Treatment Options for Varroa Mites

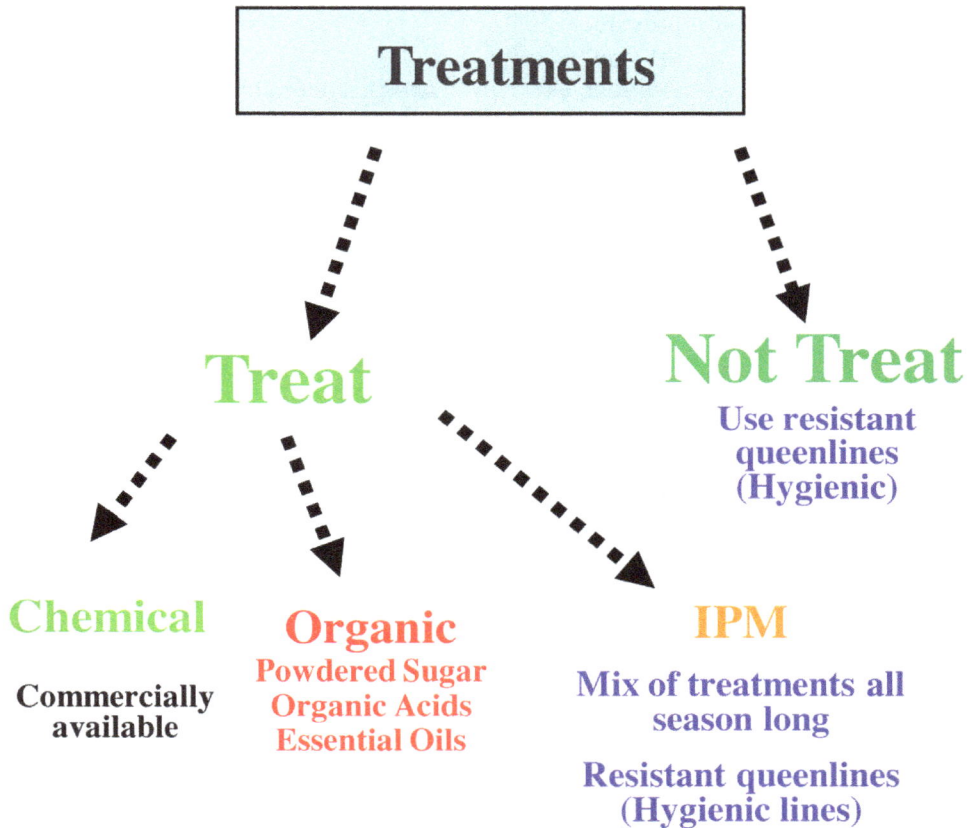

Treatments

Treat

Not Treat
Use resistant
queenlines
(Hygienic)

Chemical

**Commercially
available**

Organic
Powdered Sugar
Organic Acids
Essential Oils

IPM

Mix of treatments all
season long

Resistant queenlines
(Hygienic lines)

Chemical Treatments for Varroa

Product Trade Name	Active Ingredient	Chemical Class
Apiguard	thymol	essential oil
Exomite Apis	thymol	essential oil
Apilife VAR	thymol, eucalyptol, menthol, camphor	essential oils
Apistan **	fluvalinate	synthetic pyrethroid
Apitol	cymiazole	iminophenyl thiazolidine derivative
Apivar **	amitraz	amadine
Bayvarol **	flumethrin	synthetic pyrethroid
Check-Mite+ **	Perizin coumaphos	organophosphate
Folbex	bromopropylate	chlorinated hydrocarbon
Sucrocide	sucrose octanoate (not effective)	sugar esters
Hivestan (alkaloid)	fenpyroximate	pyrazole
Mite-Away Quick Strips	formic acid	organic acid
generic	lactic acid	organic acid
generic	oxalic acid	organic acid

*** No Longer Effective in some areas*

Other products are also available, check bee supply companies.

Compiled from Rosenkranz et al. 2010 (see References) and http://www.maf.govt.nz/biosecurity/pests-diseases/animals/ varroa/guidelines/control-of-varroa-guide.pdf

The Problem with Treating Varroa with Chemical Miticides

▶ Mites develop resistance

▶ More and more chemicals needed to control mites

▶ Residues of chemicals end up in the wax

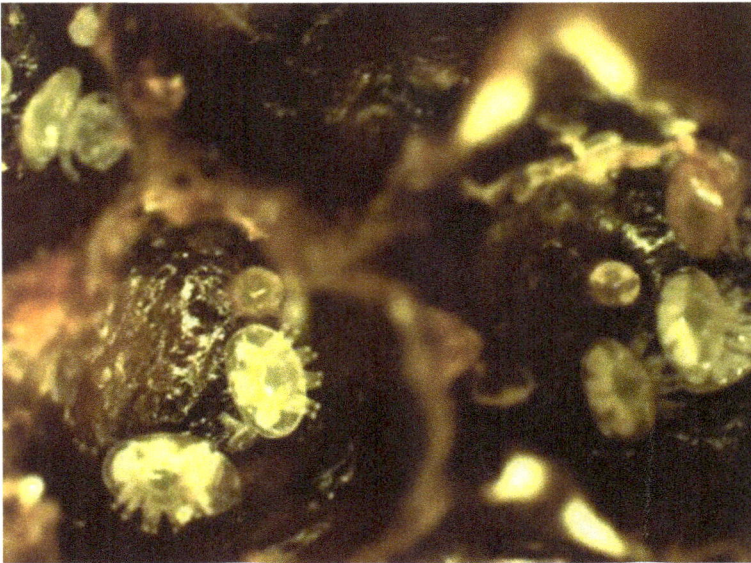

USDA photo

Consequences of Resistance

▶ Experimentation of chemical cocktails

▶ Over-application of some chemicals

▶ Higher doses used, label recommendations not followed

▶ Contamination of hive products

If selling wax products, get wax tested for pesticide residues

Residues in Hive Products

Latest findings: 121 **compounds** (mostly pesticides) found in hive products (wax and pollen)

Pollen can be contaminated if bees are near agricultural areas that are frequently treated for other pests; pollen can be trapped and discarded.

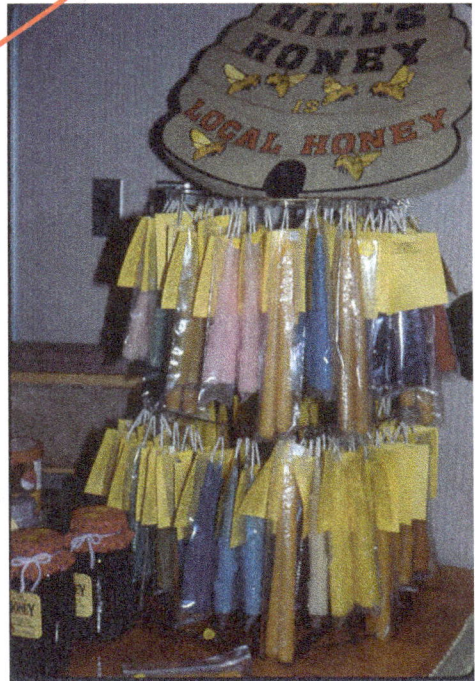

Mullin CA *et al.*, 2010 (See References)

Integrated Pest Management (IPM)

IPM uses multiple tactics to control mites

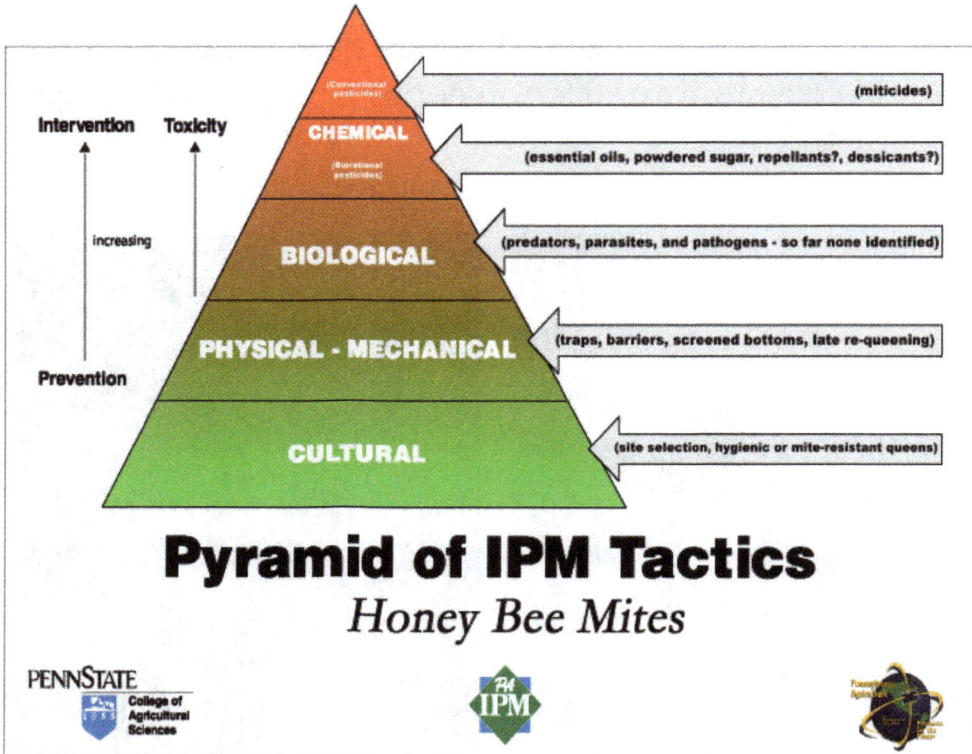

Intervention   **Toxicity**

increasing

Prevention

(Conventional pesticides)   **CHEMICAL**   (Biorational pesticides)

(miticides)

(essential oils, powdered sugar, repellants?, dessicants?)

BIOLOGICAL   (predators, parasites, and pathogens - so far none identified)

PHYSICAL - MECHANICAL   (traps, barriers, screened bottoms, late re-queening)

CULTURAL   (site selection, hygienic or mite-resistant queens)

Pyramid of IPM Tactics
Honey Bee Mites

PENNSTATE College of Agricultural Sciences

PA IPM

Queenlines Resistant to Mites; look for:

▶ Varroa Sensitive Hygiene (VHS formerly SMR)

  ✓ Bee population could be lower as pupae are pulled out to kill mites

▶ Russian queens, both mites

▶ Other lines?

  ✓ Local resistant lines: select survivor colonies and breed from them

▶ Grooming behavior, along with other traits

▶ Other Hygienic bee lines

  ✓ Some very resistant to bee pathogens as well as mites

Photo: D. Sammataro

Other Bee Mites

▶ Learn to recognize other mites

- *Acarapis woodi* or tracheal mite
  (an internal parasite)

▶ Newest Asian parasitic mites on the horizon:

- *Troplilaelaps sp.*

- *Euvarroa sp.*

Acarapis species

▶ Three mites in this Genus

✓ *A. dorsalis*

✓ *A. externus*

✓ *A. woodi*

▶ Location of these mites on a bee:

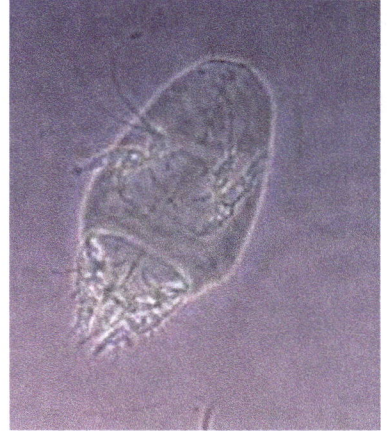

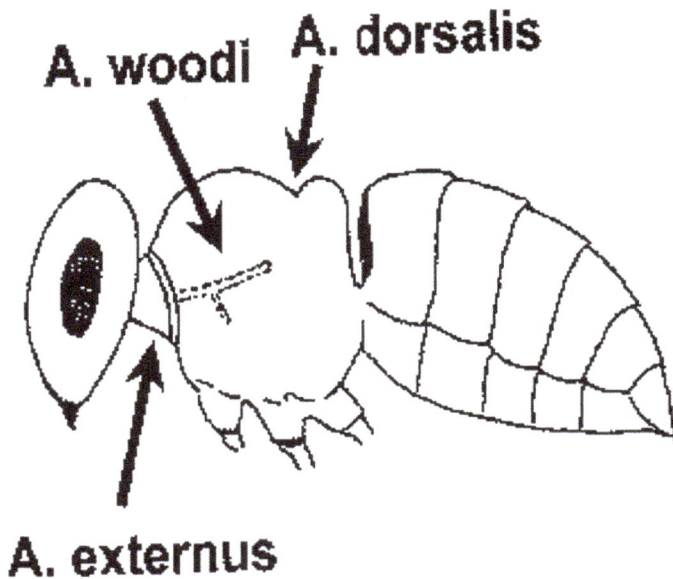

A. woodi   A. dorsalis

A. externus

L. de Guzman

Tracheal Mite

▶ The only INTERNAL bee mite

▶ Females emerge from tracheae, move off old host to quest to find another bee host.

▶ Mites attach themselves to bee hairs and move onto younger bees.

▶ Especially a problem in the winter when bees cluster.

• Mites can distinguish young bees from older bees

• Enters tube and after one to two days, lays about 1 egg/day for 12 days

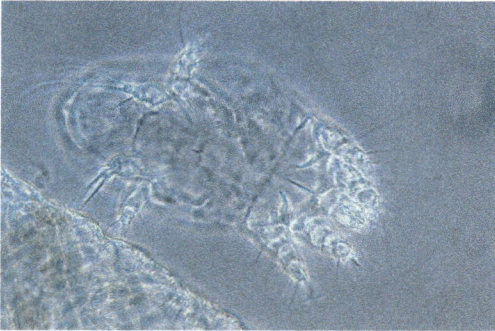

Questing Female Tracheal Mite

Drawing © by D. Sammataro 1997

Tracheal Mites

▶ Feed on bee blood (hemolymph)

▶ Attack all adults bees (especially drones and queens, which live longer than workers)

▶ Live in the respiratory system, enter through the spiracle flap on thorax

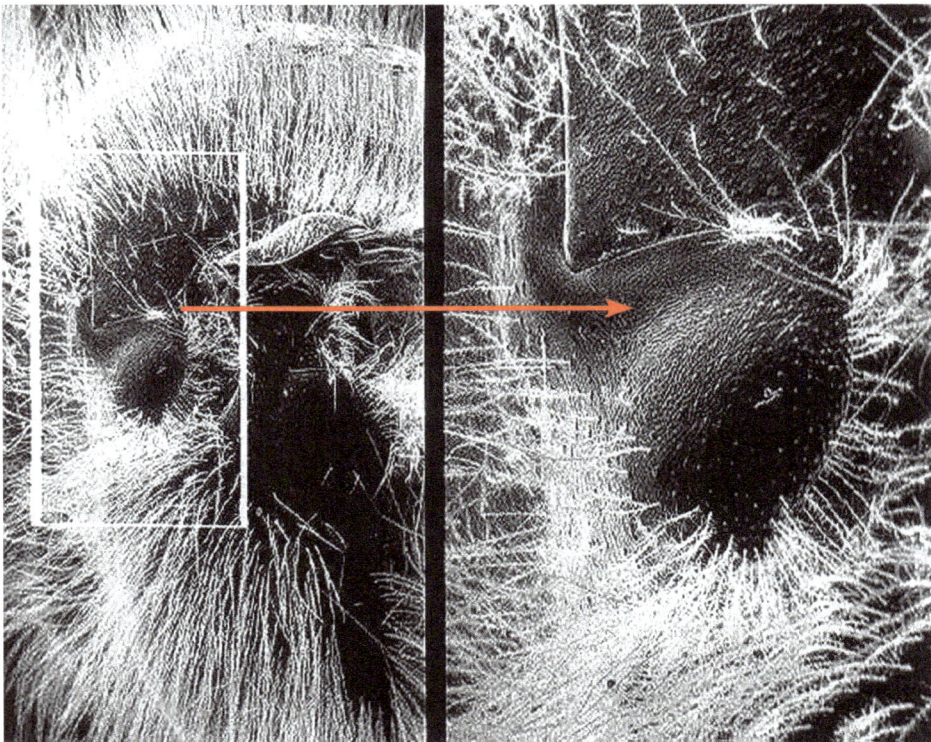

Spiracle Flap

W. Styer photos

Tracheal Mites

▶ Damage Tracheae or Breathing Tubes

- Blotches and scabs in tube
- Debris inside tube
- Puncture wounds

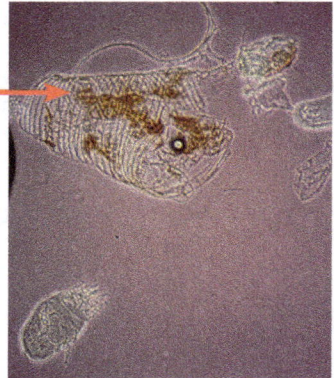

Photo: D. Sammataro

▶ Effect on Bees

- Lowers ability to use wing muscles

  ✓ Cannot keep cluster warm in winter

  ✓ Bees die over winter, crawling out of colonies

  ✓ In warmer climates, bees will survive, but viruses may be passed from mites to bees, weakening colony

  ✓ May weaken drones and queens

Tracheal Mite Dissection

▶ First Step

    ▶ Collect Bees: freeze or store in Alcohol

        • Older foragers and drones

        • Old queens

        • Older bees are on the inner cover, in honey supers or returning foragers

    ▶ Time of year is important

        • Early Spring

        • Late Fall

Bees collected using a 'bee vac'

Mite ID: Step 1

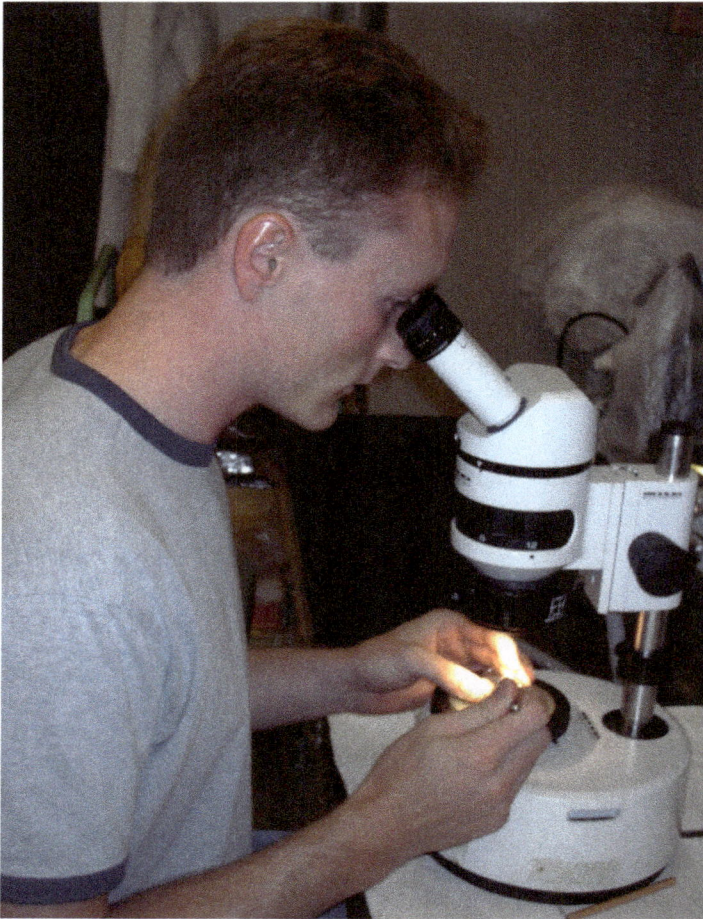

Use a Good Dissection Microscope

▶ Not Compound scope

▶ Start with the lowest power

Mite ID: Step 2

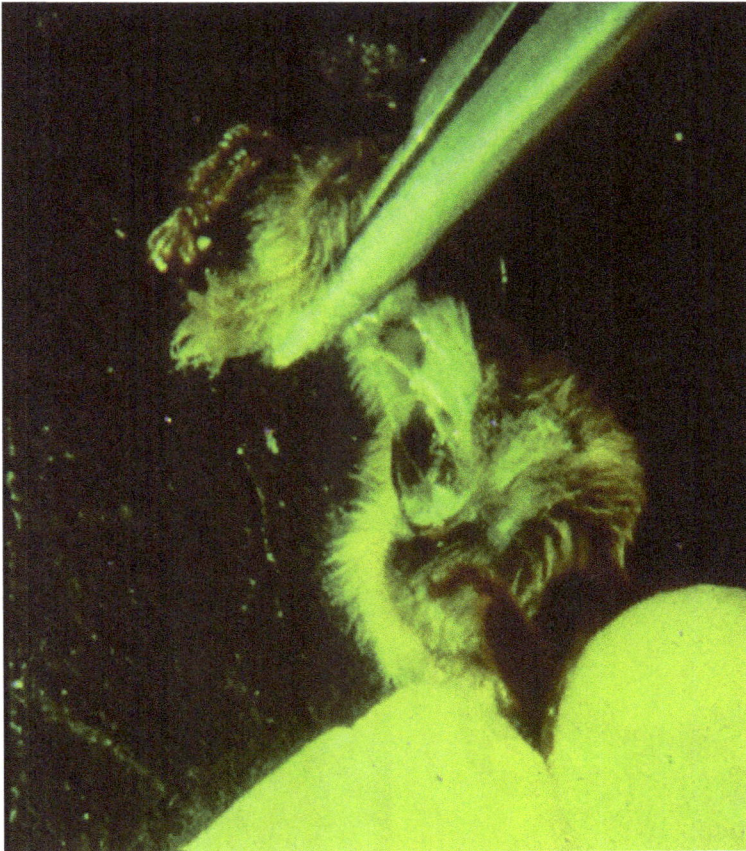

Test frozen bees (or bees stored in alcohol)

- ▶ Hold bee by thorax
- ▶ Take fine forceps and pull off head
    - ✓ First pair of legs will also come off

Mite ID: Step 3

Look through the microscope at the thorax

Branches of Main Tracheal Tubes

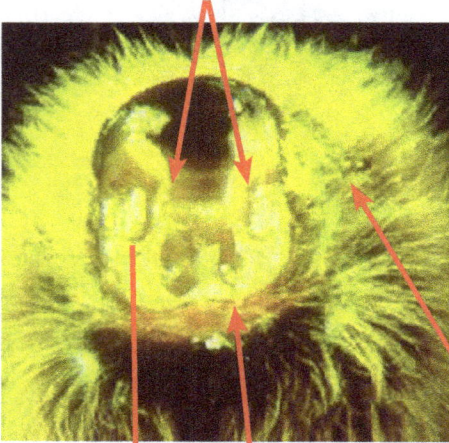

Once head is removed:

Pull off collar

You will see two tracheal branches

Spiracle Flap at Wing Base

Collar (neck of bee)

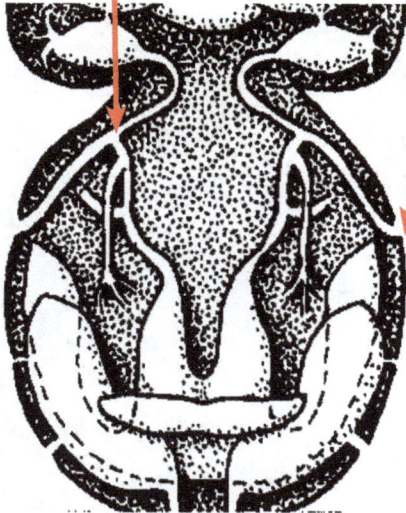

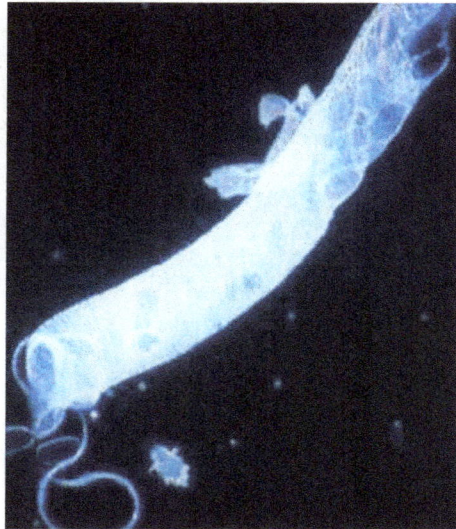

Mite-filled tracheal tube

Photos by D. Sammataro

Tracheal Mite ID

When the neck (collar) is removed:

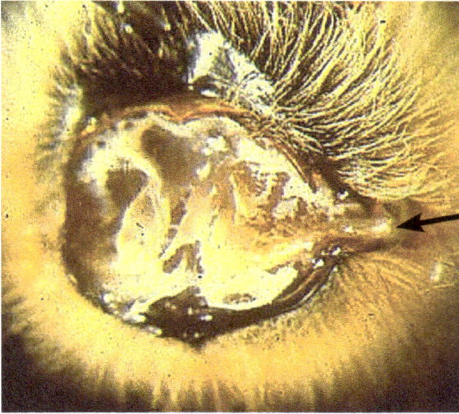

- Main tracheal tubes exposed

- Look at spiracle opening

- If infestation is light, female mite will be here

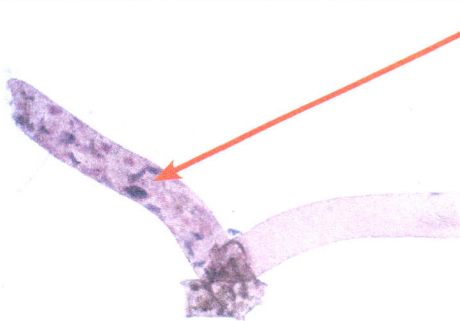

- Older infestation: many shadows will be seen

  If no shadows are visible, no mites are present

MITE DISSECTION VIDEO:

http://www.ars.usda.gov/pandp/docs.htm?docid=14370

Photos by D. Sammataro

Range of Tracheal Mites

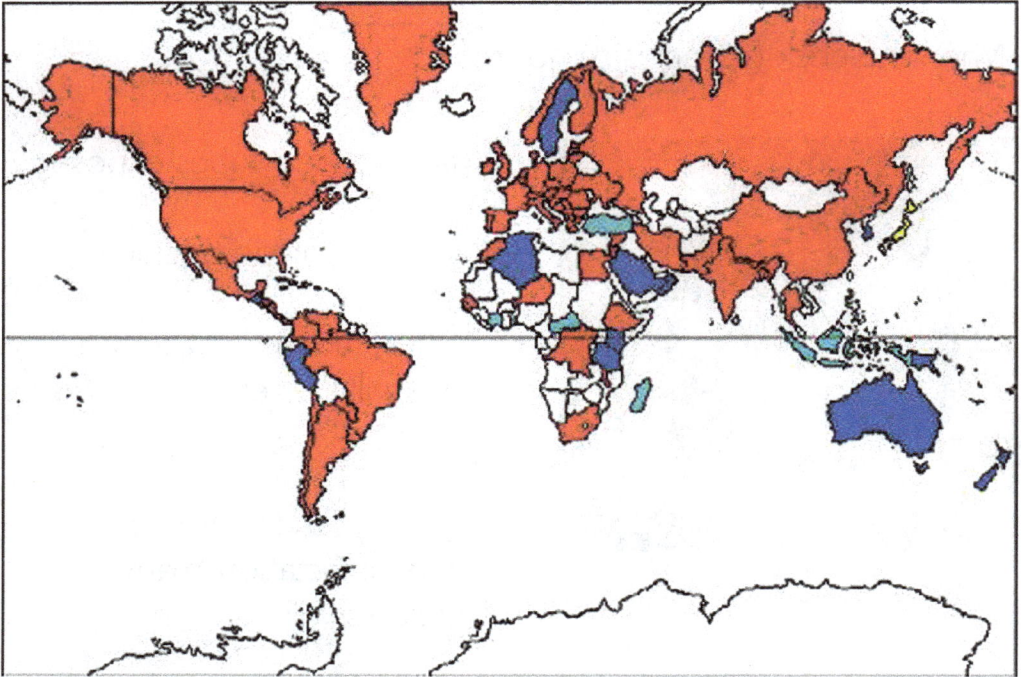

Ellis and Munn, 2005 (see References)

Tropilaelaps spp.
Next Asian Bee Mite

▶ *T. clareae* (Delfinado & Baker) found on *A. dorsata breviligula and A. mellifera* in the Philippines and Indonesia

▶ *T. koenigerum* on *A. dorsata, A. laboriosa and A. mellifera* in Asia and Indonesia

▶ *T. mercedesae* on *A. d. dorsata and A. mellifera*

▶ *T. thaii* n. sp., has been found on *A. laboriosa* in Himalayan regions

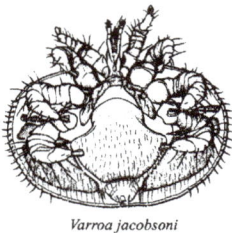

Varroa jacobsoni

Tropilaelaps clareae

T. koenigerum

T. clareae

Photos: D. Sammataro

T. clareae on *A. mellifera*

Life Cycle

▸ Similar to *Varroa* but brood is essential for survival, will not feed on adults

▸ Very serious in tropics, not so much in temperate climates (yet)

▸ Larval form is mobile, feeds actively; adults are also very fast

▸ Can carry viruses, such as DWV

Comparison of
Varroa and *Tropilaelaps*

Size comparison of *Varroa* (left) and *Tropilaelaps* (right) mites.

(Photo by Khongphinitbunjong)

Anderson, D L; Roberts, J M K, 2013 (see References)

Range of *Tropilaelaps*

▶ The known geographic range of ***Tropilalaelaps*** has spread significantly over the last 40 years. It is found primarily in Asia at this time (2014).

▶ Cold winters prevent **A. *mellifera*** from producing much brood, so any introduced ***Tropilaelaps*** will starve.

▶ Warm climates which will support uninterrupted brood production, will create continuous mite production, much like ***Varroa***.

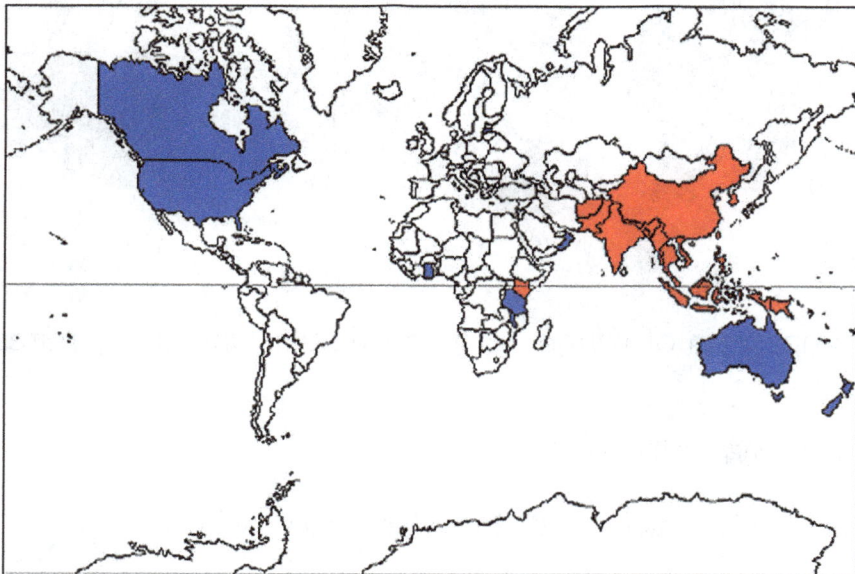

Ellis and Munn, 2005 (see References)

Control of Tropilaelaps

▶ Formic acid (70%) and Thymol

    ✓ formic acid (25 ml)

    ✓ thymol group (25g)

▶ 3.2% Oxalic Acid with 4g thymol was the best treatment for controlling these mites

▶ Also the same miticides used for *Varroa* will work on *Tropilaelaps*.

▶ Cannot survive over winter in northern climates, or if there is no brood.

    ✓ Queen can be caged to create broodless period

Another Asian mite: *Euvarroa sp.*

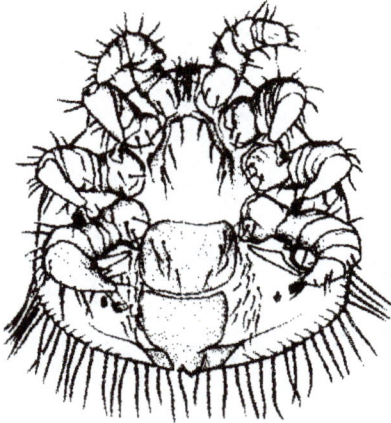

Euvarroa sinhai

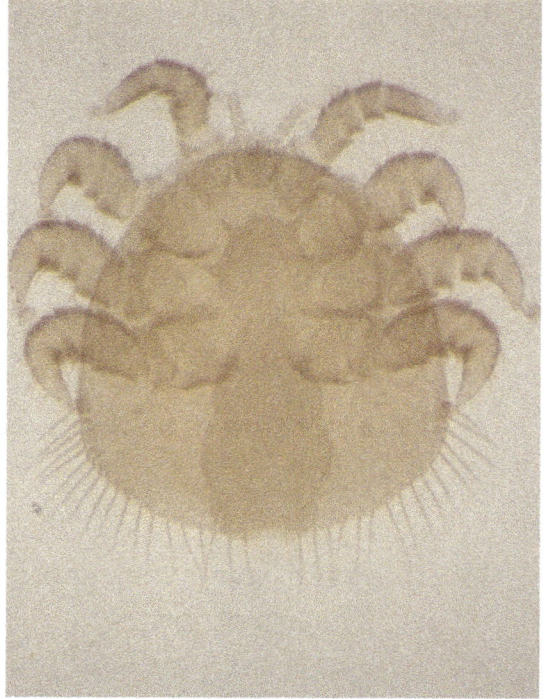

Photo by D. Sammataro

Found on the Dwarf honey bees *Apis florea* and *A. andreniformis*.

This specimen of *E. sinhai* is housed in the collection of the Ohio State University Acarology Museum.

Websites and References

Some Useful Links and References:

http://hdoa.hawaii.gov/pi/ppc/varroa-mite-information/

http://www.blackburnbeekeepers.com/Tropilaelaps(2).pdf

http://www.coloss.org/sitemap

Anderson, DL, JMK Roberts. 2013. Standard methods for *Tropilaelaps* mites research. *In:* V Dietemann; J D Ellis; P Neumann (Eds). The COLOSS BEEBOOK, Volume II, http://dx.doi.org/10.3896/IBRA.1.52.4.21

Chandler, D, Sunderland KD, Ball BV, Davidson G. 2001. Prospective biological control agents for *Varroa destructor* n. sp., an important pest of the European honey bee, *Apis mellifera. Biocont.* Sci. Technol. 11:429–448.

Delaplane, KS, JA Berry, JA Skinner, JP Parkman and AM Hood. 2005. Integrated pest management against *Varroa destructor* reduces colony mite levels and delays treatment threshold. *J. Apic. Research.* 44(4): 157–162.

Ellis, JD and P Munn. 2005. The worldwide health status of honey bees. *Bee World,* 86(4): 88–101.

Lee, KV, G Reuter and M Spivak. 2010a. Standardized sampling plan to detect Varroa density in colonies and apiaries. *Amer. Bee Journal.* 150: 1151-1155.

Lee, et al. 2010b. Practical sampling plans for Varroa destructor in *Apis mellifera* colonies and apiaries. *J. Econ. Entomology* 103(4).

Meikle, WG, Mercadier G, et al. 2008. Impact of a treatment of *Beauveria bassiana* (Deuteromycota: Hyphomycetes) on honeybee (Hymenoptera: Apidae) colony health and on Varroa mites (Acari: Varroidae). *Apidologie*. doi:10.1051/ apido:20007057.

Mullin, CA, Frazier M, Frazier JL, Ashcraft S, Simonds R, et al. 2010. High levels of miticides and agrochemicals in North American apiaries: implications for honey bee health. *PLoS ONE* 5(3): e9754. doi:10.1371/journal.pone.0009754.

Rosenkranz, P, Aumeier, P, Ziegelmann, B. 2010. Biology and control of *Varroa destructor*. *J. Invertebr. Pathol.* 103, S96-S119.

Sammataro, D. 2006. An easy dissection technique for finding tracheal mites (Acari: Tarsonemidae) in honey bees (with Video link). *International J. Acarology*, 32:339-343. http://www.ars.usda.gov/pandp/docs.htm?docid=14370

Sammataro, D and J Finley. 2004. Observations of the ectoparasitic bee mite *Varroa destructor* in honey bee (*Apis mellifera*) cells infected with chalkbrood (*Ascosphaera apis*). *J. Apic. Research*. 43 (1): 28-30.

Sammataro, D, L deGuzman, S Geroge, R Ochoa and G Otis. 2013. Standard methods for *Acarapis* research. *J. Apic. Research*. 52(4). DOI 10.3896/ IBRA.1.52.4.20 and COLLOSS BeeBook, http://www.coloss.org/beebook/II/tracheal

Webster, T and K Delaplane (eds). 2000. Mites of the Honey Bees. Dadant and Sons, Hamilton IL.